T0298906

Beyond the Energy–Water–Food Nexus

New Strategies for 21st-Century Growth

William Sarni

Author | Speaker | Advisor

will@willsarni.com

First published in 2015 by Dō Sustainability
87 Lonsdale Road, Oxford OX2 7ET, UK

Copyright © 2015 William Sarni

The moral right of the author has been asserted.

All rights reserved. No part of this publication may be reproduced, stored in a retrieval system, or transmitted, in any form or by any means, electronic, mechanical, photocopying, recording or otherwise, except as expressly permitted by law, without the prior, written permission of the publisher.

ISBN 978-1-910174-48-7 (eBook-ePub)
ISBN 978-1-910174-49-4 (eBook-PDF)
ISBN 978-1-910174-47-0 (Paperback)

A catalogue record for this title is available from the British Library.

Dō Sustainability strives for net positive social and environmental impact. See our sustainability policy at **www.dosustainability.com**.

Page design and typesetting by Alison Rayner
Cover by Becky Chilcott and Deanna Turner

For further information on Dō Sustainability, visit our website:
www.dosustainability.com

DōShorts

Dō Sustainability is the publisher of DōShorts: short, high-value business guides that distill sustainability best practice and business insights for busy, results-driven professionals. Each DōShort can be read in 90 minutes.

New and forthcoming DōShorts – stay up to date

We publish new DōShorts each month. The best way to keep up to date? Sign up to our short, monthly newsletter. Go to **www.dosustainability. com/newsletter** to sign up to the Dō Newsletter. Some of our latest and forthcoming titles include:

- *How Much Energy Does Your Building Use?*   Liz Reason
- *Lobbying for Good: How Business Advocacy Can Accelerate the Delivery of a Sustainable Economy*   Paul Monaghan & Philip Monaghan
- *Creating Employee Champions: How to Drive Business Success Through Sustainability Engagement Training*   Joanna M. Sullivan
- *Smart Engagement: Why, What, Who and How*   John Aston & Alan Knight
- *How to Produce a Sustainability Report*   Kye Gbangbola & Nicole Lawler
- *Strategic Sustainable Procurement: An Overview of Law and Best Practice for the Public and Private Sectors*   Colleen Theron & Malcolm Dowden
- *The Reputation Risk Handbook: Surviving and Thriving in the Age of Hyper-Transparency*   Andrea Bonime-Blanc
- *Business Strategy for Water Challenges: From Risk to Opportunity*   Stuart Orr and Guy Pegram

- *Accelerating Sustainability Using the 80/20 Rule*   Gareth Kane
- *The Guide to the Circular Economy: Capturing Value and Managing Material Risk*   Dustin Benton, Jonny Hazell and Julie Hill
- *PR 2.0: How Digital Media Can Help You Build a Sustainable Brand*   John Friedman
- *Valuing Natural Capital: Futureproofing Business and Finance*   Dorothy Maxwell
- *Storytelling for Sustainability: Deepening the Case for Change*   Jeff Leinaweaver
- *Beyond Certification*   Scott Poynton

Subscriptions

In addition to individual sales of our ebooks, we now offer subscriptions. Access 60+ ebooks for the price of 6 with a personal subscription to our full e-library. Institutional subscriptions are also available for your staff or students. Visit **www.dosustainability.com/books/subscriptions** or email **veruschka@dosustainability.com**

Write for us, or suggest a DōShort

Please visit **www.dosustainability.com** for our full publishing programme. If you don't find what you need, write for us! Or suggest a DōShort on our website. We look forward to hearing from you.

Abstract

WATER IS A BASIC NEED FOR BUSINESSES, the public sector and ecosystems and in a world where there is increased competition for water it is a resource at great risk of scarcity. Our growing global population and economies have created a "wicked problem" that must be considered sooner, not later, to ensure we have adequate energy, water and food in the 21st century. This concise guide goes beyond just identifying the problems and helps organizations abandon "business-as-usual" thinking to begin solving the issues found at the food, water and energy nexus. Will Sarni's framework for change for the 21st century includes both "soft path" and technology solutions and shows organizations how to engage in both paths for more complete and sustainable outcomes. Use this book to understand the connectivity between your business and the food–water–energy nexus, as well as how to begin addressing this "wicked problem."

About the Author

WILL SARNI is an internationally-recognized thought leader on water stewardship and water tech innovation and a frequent speaker for corporations, conferences and universities. He is the author of *Corporate Water Strategies* and co-author of *Water Tech: A Guide to Investment, Innovation and Business Opportunities in the Water Sector.*

Will has worked for some of the most recognized global brands to develop and implement water stewardship programs, evaluate innovative water technologies and deliver market entry strategies in the water industry. He has a creative approach in integrating diverse business and technical issues related to resource management. Will is a Board Member of the Rainforest Alliance and has worked with several enterprises as an advisor on water related programs.

He lives in Denver, Colorado.

....

THIS BOOK IS DEDICATED TO MY FAMILY FOR THEIR EXCEPTIONAL SUPPORT AND TO MY COLLEAGUES ENGAGED IN SOLVING THE CHALLENGE OF THE ENERGY–WATER–FOOD NEXUS TO ENSURE ACCESS TO THESE RESOURCES FOR THOSE IN NEED.

....

Contents

CONTENTS

Introduction

"No problem can be solved from the same level of consciousness that created it." ALBERT EINSTEIN

WHEN RESOURCES ARE SCARCE – as they are now, and will become ever scarcer in years to come – "silo thinking" is no longer an option. For too long we have managed energy, water and food as if there was only a casual relationship between them. While there is now a growing recognition that we must manage these resources as a system and that we must reduce the impact of our use of them, we have a long way to go to provide them *sustainably* and *resiliently*.

These three resources are readily accessible in developed economies while scarce and relatively expensive in emerging economies. With the majority of the world's population growth expected to come this century in emerging economies, this poses an obvious problem. In a world where the global population is headed to 9 billion in this century the question is clear: how will we deliver these resources to all in a fair and equitable manner?

The energy–water–food nexus, then, is a "wicked problem."

I was introduced to the notion of a "wicked problem" by Tom Higley, a friend, entrepreneur and founder of 10.10.10 (**www.101010.net**). When Tom asked if water was a "wicked problem," I assumed he was asking whether the "water crisis" was a difficult challenge to solve. What I didn't know was that "wicked problems" (**http://en.wikipedia.org/wiki/Wicked_problem**) actually have a specific definition.

INTRODUCTION

A 2007 report from the Australian Public Service Commission[1] outlined the characteristics of a wicked problem:

- Wicked problems are difficult to clearly define.

- Wicked problems have many interdependencies and are often multi-causal.

- Attempts to address wicked problems often lead to unforeseen consequences.

- Wicked problems are often not stable.

- Wicked problems usually have no clear solution.

- Wicked problems are socially complex.

- Wicked problems hardly ever sit conveniently within the responsibility of any one organization.

- Wicked problems involve changing behavior.

- Some wicked problems are characterized by chronic policy failure.

As I read the articles that Tom passed along, I became even more convinced the water crisis, and in turn, the energy–water–food nexus, could be framed as a "wicked problem." Many, if not all, of these characteristics fit the energy–water–food nexus challenge.

We need wickedly smart solutions for wicked problems.

Old thinking – and in many instances, old technologies – won't get us to where we need to be to provide water, energy and food to an increasing global population. Fortunately, opportunities abound in innovative thinking and technologies to address this problem.

This book developed from my focus on the impact of water-related risks to business growth and how water risks are driving opportunities for new products and services. Water sits at the intersection of the energy and food nexus – without water we can't grow crops or produce energy (renewable energy sources, with their low water footprint, are the exception to this rule). In addition, water is essential for ecosystems. While I will not directly address water's role in and importance to ecosystems, we must be mindful that any public policy and allocation frameworks need to collectively address energy, water, food *and* ecosystems. This is even more critical in an era of increased focus on the value of natural capital to both the private and public sectors.[2]

This book offers a systematic review of the trends creating stress at the energy–water–food nexus: urban growth, economic development in developing countries, population growth and climate change. It also takes a look at what is not working, what needs to change and examples of successful policies, frameworks and technologies that are addressing the nexus and building a more resilient and sustainable world. It is by no means an exhaustive treatment of the topic. Instead it is intended to show where we need to go to tackle the problems in the nexus.

The title "21st-Century Growth" refers to the fact that any successful solution to these nexus stresses must address how we can continue economic growth while also improving the quality of life in the face of ever-increasing demand for resources.

Twentieth-century thinking assumed that energy and water would forever be plentiful and cheap, and that the current agricultural practices that are collectively called the "green revolution" could always feed the world. The historical accomplishments in energy generation, water supply and distribution and agricultural productivity should not be discounted. However, the world has changed and the mindset of abundant supply is mostly gone, replaced with an awareness of resource constraints limiting economic growth and the rapidly appearing impacts of climate change.

The process of shifting from a 20th-century mindset to 21st-century thinking will be challenging but is within our reach. And it will necessarily require some tradeoffs in policy, technology and investments, among others.

The upside to these challenges is that they provide a spur to innovation in technology, partnerships and public policy. There are no technology or public policy magic bullets. Creating and implementing "soft" public policy solutions and "hard" technology solutions will not be easy, but they are essential to providing energy, water and food to the world and to raising the standard of living (access to clean water, sanitation and hygiene) for the billions who live at the base of the pyramid.

Resolutely accepting the "business as usual" scenarios of resource scarcity and decline in well-being for a large portion of the global population is a recipe for failure, despair and tragedy. Creating collective action frameworks that convene stakeholders who are otherwise siloed, and investing in technological solutions, will instead provide hope for a different trajectory.

I don't believe in business as usual. Instead I believe innovation will save us – through collective action partnerships and technology.

....

CHAPTER 1

What is the Nexus? Meeting the Energy, Water and Food Needs of 9 Billion People

THE ENERGY-WATER-FOOD NEXUS is really a collision of systems creating a more complex set of relationships, challenges and opportunities (Figure 1). Each of these systems is complex on its own, and the linkages between them make the nexus significantly more intricate still.

The food system consists of the activities, resources and people involved in bringing food from the farm to the table, including but not limited to the following:

- Growing and harvesting crops.

- Breeding, housing, feeding and slaughtering animals for food.

- Catching and harvesting aquatic plants and animals for food.

- Processing raw plant and animal materials into retail products.

- Transporting feed, animals, produce and other goods.

- Storing and selling products at retail outlets.

- Preparing and eating food.

- The land, labor, soil, energy, animals, seeds and other resources involved in making the aforementioned activities possible.

The water system provides water for human use as well as for ecosystems as a whole. Water is a local issue, with the watershed or catchment the local "unit." Watersheds are numerous: according to the US Geological Survey

FIGURE 1. The energy–water–food nexus.

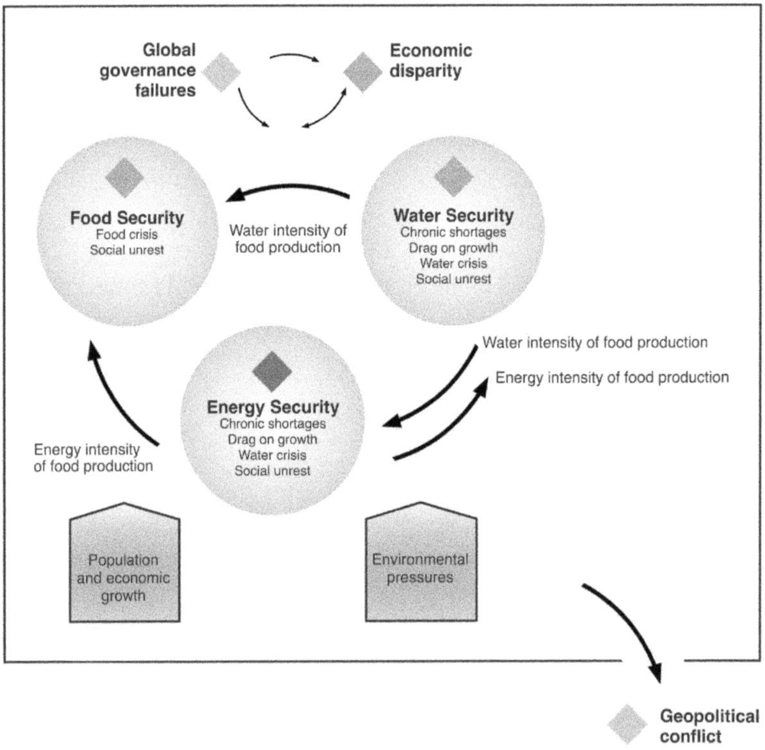

SOURCE: World Economic Forum, 2011.

(USGS) there are over 2,264 watersheds in the continental United States alone.[3] This water is often moved around within and between watersheds in order to meet the needs of individuals and communities; and in some cases we dramatically alter water systems to better meet our agricultural, municipal, commercial, industrial and energy production needs.

The overarching energy system includes not only the systems required to generate, transmit and distribute electricity, but also the systems needed to produce and distribute transportation fuels. Electricity is generated either at a power plant fueled by fossil fuels or nuclear fission, or by lower-impact sources, including hydropower, wind, solar and geothermal. The aspects of the energy system involved in producing transportation fuels include producing, refining and distributing oil and natural gas, as well as producing and processing feedstocks for biofuels, for example the maize used to produce ethanol.

As noted above, while each of these systems is complex on its own, the interactions among them are where the larger stresses arise.

Food and water

Food production and processing is an immense source of water consumption, with crop irrigation alone accounting for about 40% of all of the water withdrawals in the United States (and in states such as California as much as 80% cent of water use).[4] Irrigation competes with other major water uses such as manufacturing, power plant cooling, municipal drinking water and fossil fuel production. These water resources can be additionally strained during droughts, causing problems for farmers who rely on irrigation of their crops. Agricultural water use can also negatively affect watersheds through runoff from fertilizers, pesticides and manure from farms and feedlots polluting local water resources.

Water and energy

Generating energy from fossil fuels requires large amounts of water, primarily for cooling – although once-through generation processes actually consume only small amounts, recycling water through cooling towers for reuse. Nearly half of all water withdrawals in the United States are used for power plant cooling.[5] The hundreds of large-scale power plants across the United States together withdraw 58 billion gallons of water from the ocean and 143 billion gallons of freshwater each day.[6] This dependence is why most power plants are located near rivers, lakes or the ocean.

As with food and water, droughts and other water shortages also affect power plants. When surface water levels drop, they can lose their access to cooling water and have to reduce or shut down operations. When the weather is warmer or drier, the bodies of water that supply power plants may face temperature increases that hamper cooling processes or harm the ecology of that water system.

Food and energy

The connection between food and water is clear, but the ties between food and energy became much stronger with the "green revolution" in farming between the 1940s and 1970s. The technology and industrialization underlying the green revolution have increased the energy needs of farming and food production along the entire process of getting food from farm to table. Among the energy-intensive activities of modern agriculture are fertilizer production, water pumping, farm equipment operation, food processing and packaging and food and livestock transportation.

Population growth and associated problems are increasing the stresses upon each of the elements of the energy–water–food nexus, as well as the connections between them. Statistics compiled by the United Nations indicate that these stresses are only going to increase as the global population grows:[7]

- The global population is expected to increase to about 9.1 billion by 2050.

- The average growth rate per year from 2006 to 2050 is projected to be 0.75%.

- In sub-Saharan Africa and Near East Africa growth is projected to be respectively 1.92% and 1.19% per year.

CHAPTER 2

Global Trends in Energy, Water and Food

THE DEMANDS FOR WATER, ENERGY AND FOOD are growing unabated. To understand the solutions to the threats to the energy-water-food nexus we need to first understand these global demands in greater detail.

2.1 Energy

The demand for energy has been increasing and is projected to continue for the foreseeable future. This should come as no surprise, since the need for energy is driven by the same factors that are driving the demand for water and food. Like water and food, the energy sector is being transformed through technology innovation, led to a large degree by a need to meet demand and to reduce greenhouse gas emissions from fossil fuel energy sources.

As with water and food, population and income growth are the key factors behind growing demand for energy. By 2030, world population is projected to reach 8.3 billion, which means an additional 1.3 billion people will need energy; and world income in 2030 is expected to be roughly double the 2011 level in real terms. In response, global primary energy consumption is projected to grow by 1.6% per annum (p.a.) from 2011 to 2030, adding 36% to global consumption by 2030. The growth rate itself declines, from 2.5% p.a. from 2000 to 2010, to 2.1% p.a. for 2010 to 2020, and 1.3% p.a. from 2020 to 2030.[8]

Demand for energy will not be evenly distributed. Low- and medium-income economies outside the Organization for Economic Co-operation and Development (OECD) account for over 90% of population growth to 2030, and as a result of their rapid industrialization, urbanization and motorization, these economies will also contribute 70% of the global GDP growth and over 90% of the growth in global energy demand (Figure 2).

FIGURE 2. Projected energy demand growth by region, industry sector and fuel type to 2030, in tons of oil equivalent (toe).[8]

Industrialization and growing power demand

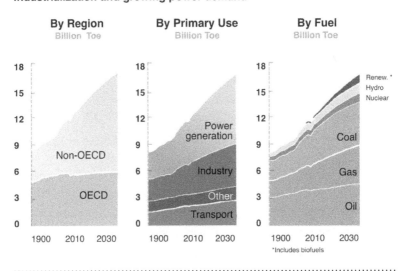

Some key takeaways from this chart:

- Almost all (93%) of the growth in energy consumption will be in non-OECD countries, with energy consumption in 2030 61% above the 2011 level, growing by an average of 2.5% p.a. (or 1.5% p.a. per capita).

- Non-OECD countries account for 65% of world consumption in 2030, compared to 53% in 2011. OECD energy consumption in 2030 is just 6% higher than in 2011 (0.3 energy – water – food % p.a.), and will decline by about 0.2% p.a. per capita from 2011 to 2030.

- Energy used for power generation will grow by 49% (2.1% p.a.) from 2011 to 2030, and will account for 57% of global primary energy growth. The primary energy used directly in industry will grow by 31% (1.4% p.a.), accounting for 25% of the growth of primary energy consumption.

- The fastest growing fuels will be renewable (including biofuels), averaging 7.6% p.a. 2011 to 2030. Nuclear and hydro both grow faster than total energy (2.6 and 2.0% p.a., respectively). Among fossil fuels, gas grows the fastest (2.0% p.a.), followed by coal (1.2% p.a.), and oil (0.8% p.a.).

2.2 Water

As mentioned above, almost 1 billion people don't have access to safe water and over 2.5 billion don't have access to sanitation and hygiene. Why is it that there are so many globally that don't have access to safe water and sanitation?

Let's first examine the issue of water scarcity and access to safe water. JP Morgan Global Equity Research framed the reasons for water scarcity in a report titled "Watching Water" published in 2008:

- **Population growth and increasing food needs (the rise of the middle class).** The current global population recently crossed 7 billion (at the time of the JP Morgan report it was about 6.4 billion) and is increasing by about 70 million people per year, with most of the growth in emerging economies. The global population is expected to reach 8.1 billion by 2030 and 8.9 billion by 2050. While growth in OECD countries is expected to remain relatively flat, the population of the United States is expected to increase from 320 million at the end of 2014 to 370 million by 2030.

- **Urbanization.** More than half of the global population now lives in cities, and increasing urbanization results in increased industrialization and increased water use.

- **Climate change.** Climate change will alter hydrologic cycles on both a regional and local level. The long-term and short-term availability of freshwater will be altered along with changes in water quality, for instance water temperature, increased dissolved constituents, and others.

A 2030 Water Resources Group (WRG) report, "Watching Our Water Future," also provided a view of water scarcity, globally and within selected regions.

WRG concluded that "there is little indication that, left to its own devices, the water sector will come to a sustainable, cost-effective solution to meet the growing water requirements implied by economic and population growth." This is not an encouraging prognosis for the future considering the increasing demand for water by both the public and private sectors.

The report makes the key points that, in the world of water resources, economic data are insufficient, management is often opaque and stakeholders are insufficiently linked. *This also sums up the challenge of the energy, water and food nexus.*

WRG lays out scenarios for water supply, water demand and the "water gap" on a regional scale. This gap will play a critical role in how businesses address the risk and opportunities for them. The report concludes that by 2030, assuming an average growth scenario and if no efficiency gains are realized, global water requirements will grow from 4,500 billion m^3 to 6,900 billion m^3 – about 40% above current accessible and reliable supplies.

The 40% gap is driven by global economic growth and development. Agriculture makes up the majority of this global water demand with current use at about 71% of total demand. By 2030, the WRG expects that agriculture's total water use will increase, but with faster population growth its share will decline slightly to 65% of total demand. Industrial demand is currently 16%, with a projected increase to 22% by 2030. Domestic water demand will decrease slightly from 14% to 12% by 2030.

The key concern in the WRG forecast is how to close the projected gap between business as usual and estimated increases in supply and water efficiency. For example, historical improvements in water efficiency in agriculture reveal only about 1% improvement between 1990 and 2004. There has been a similar rate of improvement in the industrial sector. If we project these rates of efficiency improvements to 2030, we would only meet about 20% of this 40% gap. If we assume a 20% increase in supply we would still have a remaining 60% of demand unmet.

A few of the conclusions the WRG outlines in its report:

- Agricultural productivity is a fundamental part of the solution to closing the water gap since the agricultural sector makes the greatest demands on global water use and water efficiency is one of the key low-cost technology solutions.

- Industrial and municipal productivity is just as critical as agricultural productivity improvements.

- There is a link between quality and quantity of water.

- Most solutions require cross-sector tradeoffs such as increased irrigation to promote agricultural productivity and resultant increases in energy use.

2.3 **Food**

No other resource feels the squeeze from water scarcity and quality and energy needs like food production. The world has made significant progress in raising food consumption per person, in terms of kcal/person/day – the key variable used for measuring and evaluating the evolution of the world food needs. In the last three and a half decades, consumption increased from an average of 2,370 kcal/person/day to 2,770 kcal/person/day. This growth was accompanied by significant structural changes, as diets shifted toward more livestock products and vegetable oils, and away from staples such as roots and tubers.[9]

These increases in population and economic growth will drive dietary changes. For example, total food consumption globally, as measured in kcal/person/day, is projected to increase from 2,373 in 1969/1971 to 3,070 by 2050. In developing countries the growth is projected to be from 2,056 in 1969/1971 to 2,572 in 2050. Below are some of the current and projected changes in diets.

- **Cereals** are currently the most important source of total food consumption in developing countries (their direct food consumption provides 53% of total calories) and the world as a whole (49%).

- **Roots, tubers and plantains** have traditionally been the mainstay of food consumption in several countries with low to middle levels of overall food consumption, mainly in sub-Saharan Africa and Latin America. Ten countries depend on these products for over 30% of food consumption in terms of calories and another six for 20 to 30%, all 16 in sub-Saharan Africa.

- World average **sugar** consumption per capita has been nearly constant over several decades, but rising in the developing countries and falling in the developed countries.

- The other major commodity group with very high consumption growth in the developing countries has been **vegetable oils**. The rapid growth in consumption, in combination with the high calorie content of oils and other oil-crop products, have been instrumental in bringing about the increases in apparent food consumption (in kcal/person/day) of the developing countries that characterized past progress in achieving food security.

- The growth of world **milk** production and consumption has been less buoyant than that of meat. World per capita consumption is currently 83 kg per year, up from 77 kg 30 years ago. All of the increase in per capita consumption came from the developing countries (from 37 kg to 52 kg), with China playing a major role in the last few years.

CHAPTER 3

How Did We Get Here?
What is Not Working?

NOW THAT WE HAVE A FEEL for the major drivers in water, food and energy, the way to get to solutions to the nexus is by first understanding how we got here. And "here" is not a good place to be, facing increased scarcity and stress on resources, as well as an inability to meet the needs of a growing global population.

3.1 Silo thinking

When there is a real or perceived abundance there are few, if any, incentives to think beyond the silo in which you operate. If you are in the water industry and supplies are abundant, then there is no need to understand how the energy and food sector may need the water you are using. Same with energy: if you are in the power generation or oil and gas sectors there is no need to understand demands from the agricultural sector. Historically, this has often been the case.

This is now changing.

As a result of increased competition for resources, coupled with innovation driven by scarcity and an awareness of sustainability and resilience, the silos are breaking down.

Two energy companies offer good examples of moving beyond silo thinking within the power generation sector: EDF and Southern Company.

The EDF Group, one of the leaders in the energy market in Europe, is an integrated energy company, active in generation, transmission, distribution, energy supply and trading. For many years, the EDF Group has been deeply involved in water issues: water resources forecast and management, water quality, fish migration, sediment transportation.

In order to optimize and balance water allocation between energy generation and water's many other uses in France's Durance Valle – including agriculture, tourism, flood control and drinking water – as well as to prepare for future water demand from these and other uses, EDF implemented an innovative solution.

In 2000, the company signed a six-year Water Saving Convention along with the two main irrigation users in the valley, with a goal of saving 44 million m^3 of water, with EDF offering financial incentives to reach the target. The saved water could then be used to generate additional energy during peak demand times.

The program was so successful that in 2003 the participants raised the savings target to 65 million m^3, and then in 2006 to 90 million m^3. As a result, annual agricultural water consumption decreased from 310 million m^3 in 1997 to 201 million m^3 in 2006.[10]

The Atlanta-based **Southern Company** is the largest energy company serving the southeastern United States, with 4.4 million customers and nearly 46,000 megawatts of generating capacity. The company's Environmental Management Council oversees Southern's compliance with environmental standards and develops its environmental policy and action plans.

Southern's Plant Crist electric generating plant on Florida's panhandle uses water as a cooling agent in the steam production process to operate the advanced scrubber system that reduces air emissions. As such, the plant needed an innovative way to efficiently cool the scrubbing system while saving scarce water resources. The company partnered with Emerald Coast Utility Authority (ECUA) to develop a sustainable system to incorporate treated wastewater into the electric generation process, saving billions of gallons of water.

By using water from a nearby ECUA treatment plant, Southern is able to avoid taking millions of gallons every day from the nearby Escambia River and achieve zero discharge. The Southern Company now has 16 power plants with the same type of once-through cooling, which withdraw and return approximately 4 billion gallons per day from lakes, rivers and estuaries.

3.2 Public policy

Historically, public policy has been crafted through the lens of abundance. When resources abound, there is little to no need for conservation, sustainability or pricing according to value.

This is also changing in response to water scarcity. The states of Texas and California are examples of places where public policy is evolving to address the realities of water scarcity and a move toward low-carbon power by adopting solar and wind energy. Even agriculture is adapting to less water by moving to water-efficient irrigation systems. Over the past two years, both of these states have approved increased funding to address water scarcity and the resultant impact on their economies.

One of the major initiatives to address the nexus through changes in policy is the work coming out of the Bonn 2011 conference on "The Water, Energy, Food Security Nexus, Solutions for a Green Economy." The Bonn conference served as the major input from the German Government to Rio+20, the United Nations Conference on Sustainable Development in Rio de Janeiro in June 2012, 20 years after the 1992 Earth Summit. The output from Bonn 2011 gave the nexus a prominence it hadn't previously achieved, and provides an excellent overview of the policy shifts that need to occur and a framework for thinking about nexus solutions. It framed an approach to optimize resource use, balance allocation between competing uses and stimulate economic growth. The goals were to "contribute to eradicating poverty; ensuring water, energy and food security for all; achieving sustainable and equitable development; maintaining productive and resilient ecosystems; and reducing vulnerability to climate variability and change." The conference participants also identified three "action fields" reflecting the three dimensions of sustainable development – economic, environmental and social.

- Accelerating access, integrating the bottom of the pyramid: meeting minimum standards of access to safe water, adequate sanitation, healthy food and clean sustainable energy are prerequisites for human development and dignity.

- Creating more with less: there is a growing recognition that reducing waste, limiting over-use and increasing overall economic productivity is not only essential as demand on the world's resources increases, but also makes sound economic and business sense.

- Investing to sustain ecosystem services: the contribution of ecosystems to human well-being and the economy is considerable. The services provided by ecosystems include freshwater, food, including crops, fish and other aquatic products, timber and fiber production, biofuels, climate regulation and tourism.

Bonn 2011 used these three action fields to better understand the inter-linkages between water, energy and food security, to discuss conditions for the transition to a greener economy and to identify the coordinated policy environment and incentives to trigger change. It also made the following recommendations concerning policy:

- Benefit from policy coherence: ensure the interdependency between water, energy and food security is explicitly identified in decision-making within and across all levels to realize the potential for mutually beneficial action and avoid conflicting policy objectives and unintended consequences.

- Unlock productive potential: overcome constraints to bridge the gap in provision of basic services, use innovation and efficiency gains to provide economic returns and optimize allocation of scarce resources for productive use.

- Reduce and re-use in multiuse systems: promote changes in the way waste is viewed, to minimize its creation, raise levels of re-use and recycling and ultimately to see waste as a resource that can be managed for multiple uses, including increasing the availability of food, for energy generation and as a source of water for drinking, industry and the environment.

- Gain from healthy and productive natural assets: recognize the considerable multi-functional value of natural assets and ecosystems to provide food and energy, conserve water, support livelihoods and contribute to the economy and appreciate the role that biodiversity plays in supporting life and cultures.

• Mobilize consumer and public influence: acknowledge the catalyzing role that individuals have in changing consumption patterns, reducing water and energy footprints and improving efficiency of resource use both through their direct actions as responsible members of a fair and just society and in influencing the way business is done.

Each of these represents a logical area where policy needs to change to reflect the reality of increased competition for resources and the need to provide services to the increasing population and ecosystems.

3.3 Market failure

As with wicked problems – complex challenges with multiple distinguishing features – market failures occur when one or more of the following situations exist: no capital is being spent to address a problem; capital is being spent but without the desired result; no capital is being spent because nobody knows it's a problem; the problem is known but no one can imagine that it's not already being addressed; or no one is addressing it because a solution is thought to be impossible.

Externalities are a type of market failure and water is one such externality. David Zilberman of the University of California has defined an externality as existing when the welfare of some "agent, or group of agents, depends on an activity under the control of another agent. Under these circumstances, an externality arises when the effect of one economic agent on another is not taken into account by normal market behavior."[11]

In other words, when a resource that one group depends on is controlled by another group, market forces are insufficient to encourage responsible use of that resource.

When an externality exists, the prices in a market do not reflect the true marginal costs and/or marginal benefits associated with the goods and services in the market. Individuals acting in their own self-interest will not have the correct incentives to maximize total surplus. Competitive markets are inefficient when externalities are present, and so governments often take policy action in an attempt to correct or internalize externalities. The use of water in energy and food production would be considered a "production externality." In this case, the impact of food or energy production on other uses of water – such as ecosystems – would not be considered. If food or energy production had to account (pay) for downstream impacts, then they would likely change their behavior.

Within the energy–water–food nexus, two examples illustrate market failures and their impacts: water pumping in Punjab, India, and global biofuels production.

3.3.1 Punjab, India: Free power has led to unabated pumping of water resulting in groundwater depletion

In the past half-century, India has undergone a successful transformation from a nation suffering recurring famines to one that produces a surplus of grain for its population. This shift has occurred in large part through its rapid expansion of irrigation, which increasingly comes from groundwater.[12] This success has also come at a cost, as rapid depletion of India's groundwater threatens further growth, particularly in the state of Punjab, the "breadbasket" of India.

As practices of the agricultural "green revolution" took root in Punjab, the Indian government focused on the state as a source of rice and wheat for the national food procurement and distribution system, offering a guaranteed price to farmers. Across the state, governments at every level created incentives to provide farmers with cheap institutional credit, as well as other subsidies to encourage private investment in irrigation, including tanks, wells, pump-sets and irrigation structures. Governments also subsidized energy used for irrigation, resulting in a rapid expansion of irrigation wells and pumping operations. Over the course of 30 years, rice and wheat production in Punjab grew three-fold as productivity and the number of operations increased.

But this cheap or free electricity has led farmers to over-irrigate their crops by using cheap, inefficient, poor-quality motors – resulting in high energy use, depletion of groundwater resources and a drop in water tables. Further, the creation of a minimum support price coupled with cheap electricity resulted in a water-intensive annual rice–wheat crop rotation that consumes much more water than the annual rainfall. As irrigation has shifted from surface canals to pumping groundwater, energy use for farming has increased alongside water use, and together with the annual rice–wheat crop rotation, has resulted in Punjab's transformation from a water-surplus to a water-scarce state.

3.3.2 Biofuels: Has global production increased the cost of food?

As biofuels are produced from crops – including food crops – there is persistent concern around the globe, particularly in regions that suffer from food insecurity, that biofuels compete with food production and are responsible for increases in food prices. The assumption is that this competition results in more food insecurity. Many people accept this simple causality chain.

Global biofuels production accounts for a significant part of global use of a number of crops. Data from the UN Food and Agriculture Organization (FAO) and the OECD state that between 2007 and 2009, 20% of all sugarcane, 9% of both oilseeds and coarse grains and 4% of sugar beet were used for biofuels.

It is unclear where the world will land with respect to biofuels and food. While there are concerns that public policies on biofuels can impact food production, there remain policies in place to promote the use of biofuels. For example, in the European Union, the Renewable Energy Directive requires a 10% share of renewable energy in the transport fuel mix by 2020, subject to the sustainability of production and commercial viability of second-generation biofuels.[13] These targets are for renewable energy in any form, but current technology and infrastructure mean that biofuels produced from grain and other foodstuffs are the most cost-effective way to meet them.

US biofuel policy consists of quantitative mandates for biofuel consumption requiring that by 2022, 36 billion gallons of renewable fuel be consumed annually and of this 15 billion gallons come from maize-ethanol. This amount, specified in the Renewable Fuel Standard, translates to a need for about 143 million tonnes of maize in 2022 – equivalent to 45% of the 2010–2011 maize harvest in the United States. Since the United States is the only major maize-ethanol producer, this is essentially a US production mandate. Until January 2012, there was also an ethanol blenders tax credit and import duties payable on biofuels.

....

CHAPTER 4

What Has To Change?

4.1 Decoupling growth from resource use

HISTORICALLY, business growth has been driven by increased resource use. The faster a company grows, the more resources it needs; in many cases growth happens with little attention paid to resource efficiency and reuse. The ability of companies to grow through the inefficient use of resources is becoming more limited due to increased competition for these resources and increased prices. Instead, we are seeing companies decouple business growth from resource use. This decoupling is an important start in meeting global needs for energy, water and food within our resource constraints. While this chapter is focused on decoupling in the private sector, the issue is also critical for the public sector – how to grow countries, regions and states in a sustainable way.

One company to watch when it comes to decoupling business growth from resource use – though far from the only company succeeding on this issue – is Unilever. In November 2010, Unilever launched the Unilever Sustainable Living Plan (USLP), the company's blueprint for achieving its vision to double the size of the business while reducing its environmental footprint and increasing positive social impacts. The key 2020 goals from the USLP are:

- help more than a billion people take action to improve their health and well-being

- source 100% of agricultural raw materials sustainably

- halve the environmental footprint of Unilever's products across the value chain.

Unilever's strategic approach to water includes reducing water use in agriculture and manufacturing as well as the water associated with consumer use. Two more of the company's 2020 commitments are to reduce water abstraction by its global factory network to 2008 levels or lower and to halve the water associated with the consumer use of its products in seven water-scarce countries, which represent more than half the world's population.

With regards to energy, Unilever has reduced the greenhouse gas footprint of the manufacturing and use of its products by about 6% per consumer since 2010. The company has also reduced greenhouse gas emissions from transport and manufacturing by increasing efficiency and renewables. Unilever's plan for sustainable growth both conserves ever scarcer resources around the world and provides profitable growth for the business through increased efficiency. Other companies have set similarly ambitious targets and have decoupled business growth from resource use.

4.2 Moving out of the silos – What do solutions look like?

How do we feed and provide energy and water to an increasing global population while also improving the quality of life? Fortunately, several examples of resource-stewardship best practices and innovation show how to address the stresses to the energy–water–food nexus.

4.2.1 Energy

At both the relatively small and relatively large scales, energy generation practices are starting to incorporate responsible management of water. Good examples are Browns Ferry in the United States and the international cooperation over access to water and hydropower generation in the Nile River basin.

Browns Ferry Nuclear Plant, United States

Today, the Browns Ferry Nuclear Plant is just one of 100 nuclear power plants in the United States – and one of about 435 in the world – but when Browns Ferry opened just outside Athens, Alabama, in 1974, it was a big deal. The plant was the first in the world to be able to generate more than 1 gigawatt of electricity, making it the largest nuclear power plant in the world.

Like most methods of generating energy, nuclear power relies on large amounts of water for cooling – and nuclear plants do require a tremendous amount of water.[14] Older nuclear power plants use "once-through" cooling that withdraws large quantities of water from a nearby river or lake, uses it once to cool the steam used to turn the turbines that make energy, then discharges it back to the water source. The other cooling option is a "closed-cycle" or "recirculating" system, which sends the water through cooling towers and then reuses it.

Browns Ferry now uses a hybrid cooling system, either drawing the cooling water once-through or recirculating it depending on several factors including water and air temperatures. A 2013 report from the Grace Communications Foundation profiled Browns Ferry[15] as an example of some of the challenges within the energy–water nexus. The report notes that hot summers have hindered the plant operators' ability to use its cooling towers to cool enough water to maintain operations. And because the plant operates under a state permit that requires it to maintain water temperatures at its source reservoir at or below 90 degrees Fahrenheit,[16] Browns Ferry has been forced to limit its operations at times in recent years.

During the extended drought in the Southeast during the mid-2000s, as well as during heat waves of 2007, 2010 and 2011, Browns Ferry was obligated to cut its operations even during peak demand times, requiring the plant operator, the Tennessee Valley Authority (TVA), to purchase other energy at much higher costs.

The solution? In 2010, the TVA began upgrading four of Browns Ferry's six cooling towers, as well as adding a larger seventh cooling tower. With the upgrade, the power plant is now able to run in closed-cycle mode for the first time in more than two decades. However, while the $160 million investment makes sense in the near term, if climate change brings the region more frequent and more intense heat waves and droughts, these upgrades may not be sufficient. One 2010 study[17] found that by the end of the 21st century the Southeast could be facing 60–80 more days of heat waves every year, a troubling prediction for power plants like Browns Ferry.

Regional collaboration on biofuel and hydropower development in Nile Basin of Ethiopia

Another, very different example of challenges at the energy–water nexus comes from Ethiopia, a nation where 77% of the people lack access to electricity[18] and which currently uses firewood and charcoal for non-renewable biomass sources and biomass for almost all of its energy. But with its position upstream in the Nile River, Ethiopia boasts great hydropower potential, and the country has recently made commitments to increase its energy generation from 2,000 megawatts to 10,000 megawatts per year, using as much renewable energy as possible.[19] But a report prepared in advance of the Bonn 2011 conference by the Stockholm Environment Institute[20] explores how the development of hydropower and biofuels has led to scarcity in water and land resources.

In the Awash River basin near the capital of Addis Ababa, hydropower competes with irrigation of plantations growing sugarcane and other crops for biofuels (as well as flower crops for export). Although biofuels are still a small component of global energy use, push through policies like the Renewable Fuel Standard in the United States has encouraged growth in biofuels in developing economies such as Ethiopia.[21] And because UN Food Programme statistics show that Ethiopia is far and away the world's largest recipient of food aid,[22] food security and water usage concerns related to expanded biofuels production could pose a significant perception problem if it is seen as using land and water resources to grow fuels instead of food. Egypt in particular has voiced opposition to developments upstream on the Nile that could affect river flows into the country, such as the Grand Renaissance Dam, currently under development.[23] Ethiopia's Tana-Beles Corridor is also host to large-scale biofuel plantations, with investments in recent years aiming to develop over 200,000 hectares of biofuel plantations.

The solution? In the Nile Basin, development of capacity for biofuels and hydropower provide opportunities for collaboration and regional integration by sharing the benefits associated with water and energy. Reservoirs in Ethiopia can offer benefits by reducing the siltation and higher water evaporation of downstream reservoirs in more arid and hotter climates. And Ethiopia can adopt improved agricultural technologies and make better use of its high annual rainfall to increase biomass production without affecting river water availability.

Recently there have been reports of discussions between Sudan, Egypt and Ethiopia initiating a level of trilateral consultations about new infrastructure along the Nile. As Egypt invests in land in Ethiopia, it is getting more actively involved in upstream water development in the Nile Basin.

Another framework for regional collaboration is the Nile Basin Initiative (NBI), which serves to manage transboundary tradeoffs and opportunities inherent in hydropower development, growth of agricultural markets, and regional trade opportunities. The NBI is also exploring the possibility of treating ecosystem services as regional public goods. These alignments of sector development policies with regional markets and cooperation are good examples of a nexus approach to better managing and governing limited resources.

4.2.2 Water

Ningxia region, China

The region of Ningxia in northwestern China is bordered by three of China's largest deserts and as a result has very low water availability – 200 m³ per capita per annum, or about 15% of China's average. Climate change and pollution are further shrinking water availability, with one estimate suggesting that as many as 90% of the aquifers under China's major cities are polluted.[24]

Severe droughts affecting the region since the 1990s, as well as widespread desertification, have further reduced the productivity of Ningxia's land and water. While foreign aid is supporting reforestation projects to help rehabilitate the land, those projects come at the cost of using already scarce water resources to support the new forests. At the same time, changing diets among the Chinese have also rapidly increased water demands: a 2008 study found that the annual water demand for food in China has more than tripled between 1960 and 2000, rising from 250 to 860 m^3 per capita.[25]

Irrigation water in Ningxia comes mostly from the Yellow River, but the country is expanding its long-distance water-pumping projects,[26] while also promoting less water-intensive agricultural methods, including drip irrigation, no-till farming, and choosing different crops. But water-pumping requires extensive energy use, and the region relies heavily on coal for generating electricity as well as for household heating and cooking. Because of the water intensity of generating electricity from coal – 20% of China's consumptive water use goes into coal-fired power plants – as well as coal's high carbon intensity, China is seeking and testing energy savings and alternative energy sources.

Ningxia is exploring ways to make use of its abundant solar and wind resources, as well as biogas production from pig farming and sanitation. Pilot programs run by the Ningxia Centre for Environment and Poverty Alleviation achieved a 30% reduction in household coal use.[27] While biofuels may provide new local opportunities for cleaner energy access and improved rural livelihoods, it is important to assess the overall resource-use efficiency and risks for food security of biofuel production. Due to national food security concerns China, which is the third-largest bio-ethanol producer after the United States and Brazil, has recently moved away from maize to other feedstocks, such as sweet sorghum or jatropha.

China is encouraging more efficient industrial water use by granting additional water rights to companies that install water-saving measures. There is further potential for increasing water use efficiency by allocating water to the service sector, which is generally less energy-intensive than the chemical sector.

South Africa

South Africa is a water-scarce country with low rainfall – about 50% of the world average – and one of the lowest runoffs in the world. Rainfall is also highly seasonal, with around 80% occurring within a span of five months. The country of about 50 million people faces the challenge of freshwater scarcity, which is exacerbated by growing demand, water pollution, unsustainable usage and wastage. Factors such as climate change and population growth also lead to an increase in water consumption.

In addition to having highly stressed water basins for allocation purposes, coal thermal power plants – which, like nuclear above, can be highly water-intensive – account for almost 90% of the nation's installed power capacity.[28] Competition for water across sectors is expected to increase, and the government has given power plants priority, which could negatively affect other sectors, such as agriculture. Energy companies are also exploring the possibility of hydraulic fracturing (fracking) for shale gas, which will put additional pressure on water resources.

Although the country's National Water Act (NWA) of 1998 provided an institutional blueprint for the future management of water resources,[29] it did not set a timeframe for implementing any of the water-saving innovations that the Act detailed. This was a deliberate move that allowed political heads and senior managers to determine their own priorities and implementation programs.

The first National Water Resource Strategy (NWRS) in 2004 was a step in this direction, establishing a comprehensive agenda for action. In addition to a full list of infrastructure investments, the agenda covered a range of management and institutional activities, including compulsory licensing; establishing catchment management agencies; empowering water management institutions to own, operate and maintain physical infrastructure; establishing new water user associations; and creating and expanding monitoring networks and information systems.

The NWRS identified a range of development opportunities pertaining to areas where water resources were available to promote economic activity that would create opportunities for historically disadvantaged individuals and communities. Some of the proposed measures included expanding irrigation in the Orange and Great Fish Rivers, improving irrigation schemes in some water management areas, and possible forestry development along the eastern coastal water management areas.

4.2.3 Food

Food waste management in the United States

Recent research suggests that as much as one-half of the more than 590 billion pounds of food produced each year in the United States is wasted – meaning that the nation spends about $165 billion per year to produce food that is never eaten. Just as with other industrialized nations, much of that waste takes place in homes, with the average American throwing away 20 pounds of food every month. Per capita food waste in the United States has increased by 50% since 1974.[30]

Discarded food is the largest portion of the municipal solid waste stream that ends up in landfills, with one study from the US EPA finding that less than 3% of wasted food was composted. In addition to waste, this discarded food represents a significant cost to local governments, which is why cities and towns across the United States are adopting food-waste collection and composting programs, with mandatory food recycling laws in place for businesses in the city of San Francisco,[31] for example.

Much of this food waste is a result of spoilage, over-cooking, over-purchasing and over-serving. In the United Kingdom, the recycling advocacy group WRAP has found that two-thirds of the country's food waste each year is due to food not being used in time, while the remaining third comes from people cooking or serving too much. Expiration dates are another cause of food waste, with consumer confusion about whether food is spoiled by that date or simply less than perfectly fresh resulting in still edible food ending up in the garbage.

Among food producers, food waste occurs when crops are not harvested or not sold at the market, particularly if their appearance doesn't meet supermarkets' or consumers' standards. "Sell-by" dates also result in markets discarding perfectly edible food.

Every aspect of the energy–water–food nexus is significantly impacted by all of this food waste: wasted food also leads to massive wastes in the water and energy needed to get food from farm to field, water for growing, processing, packaging, transporting and preparing foods. About 2.5% of the total energy budget of the United States is discarded as food waste[32] – the equivalent of hundreds of millions of barrels of oil. And the amount of water consumed each year to make the food that is then wasted is equivalent to Lake Erie in volume.[33]

The ecological impacts of food waste are also not to be ignored. Around the world, agriculture is converting forests, wetlands and grasslands into farmable land; reducing the amount of food waste would thus also increase the amount of land available to maintain biodiversity and continue its carbon-storing capabilities. And less food wasted would mean less pesticide and fertilizer runoff into the world's water bodies – for instance, a much smaller or nonexistent annual "dead zone" in the Gulf of Mexico caused by agricultural runoff along the Mississippi River.

Plenty of proposed and implemented solutions to the problem of food waste already exist, whether an incremental change like requiring food composting – which then can be used to grow more food on local or regional farms – or a dramatic reimagining of the food life cycle, creating a system to connect homes to their surrounding communities so that food that might be wasted in one household could instead go to another nearby home that would otherwise go hungry.

Making the food system more efficient offers environmental benefits in the form of more efficient use of resources, financial benefits of cost savings, and social benefits by reducing food insecurity. In this way, a complex problem can be addressed by individuals and groups working in every part of society.

Conservation Agriculture and sustainable crop intensification in Zimbabwe

Despite significant advancements in agricultural technology and practices, recent years have seen notable declines in agricultural productivity in Zimbabwe and many other countries in sub-Saharan Africa. Low yields, declining soil fertility, unpredictable precipitation levels and unstable markets have all caused ongoing deficits in the availability of cereals, and therefore the profitability and sustainability of smallholder farming in Zimbabwe.

Agriculture provides livelihoods for the majority of the country's rural population,[34] and more than 70% of the entire nation is indirectly or directly dependent on agriculture for employment. Two-thirds of the farms are rain-fed systems, although most of the country received low annual rainfall and as much as 50% of seasonal precipitation is lost to evaporation. Although the government has expanded the area under cultivation to try to offset reducing yields, Conservation Agriculture is increasingly embraced to reduce the negative impacts of some of the factors that are limiting agricultural productivity.

The core practices of Conservation Agriculture – minimum soil disturbance, maintaining organic ground cover and rotating crops to improve soil fertility – have already shown potential to improve Zimbabwe's agricultural production. And local research has shown that Conservation Agriculture can also reduce soil erosion and water runoff while also increasing the economic returns from farming.

The current crop of Conservation Agriculture initiatives in Zimbabwe was implemented as humanitarian programs to improve food security among communal farmers. Expanding the practices' reach to more farm communities has been slow, in part because of the perceived high labor demands of manual Conservation Agriculture practices as well as maintaining a sufficient amount of organic ground cover when farm fields are also grazed by livestock. But ongoing support for Conservation Agriculture from the government as well as philanthropic donors shows the benefits of the practice.

Farmers implementing Conservation Agriculture are seeing more intensive production on their fields, increased resilience to dry spells, and more efficient use of organic and inorganic fertilizers. Data from the UN's Food and Agriculture Organization show that, in 2012, more than 300,000 communal farmers had implemented some components of Conservation Agriculture over an area of about 100,000 hectares.

4.3 Integrated solutions in the utility sectors

There are significant opportunities for the deployment of integrated solutions in the utility sectors - energy and water. The opportunities present in the utility sector are important to highlight because they represent a shift in thinking - moving out of silo thinking.

4.3.1 "Conservation synergy"

Perhaps the most basic approach to developing holistic solutions in the energy-water-food nexus is coordination of conservation programs. A couple of reports prepared by US NGOs provide guidance and examples of the benefits of "conservation synergy."

The 2013 report, "Conservation Synergy – The Case for Integrating Water and Energy Efficiency Programs" by the Western Resource Advocates, makes the case that joint efficiency programs between energy and water utilities are a good business decision. Good business is characterized by "higher participation rates, increased customer satisfaction, coordinated and complementary program design, and an improved reputation for working smarter – not harder."

The collaboration process mapped out by the Western Resource Advocates is straightforward, bringing together top management and staff from regulatory bodies to conduct market analyses, identify the best opportunities for collaboration as well as the costs, benefits, risks and financing options for projects, then obtaining regulatory approval, conducting the projects and then repeating the cycle.

Several of the case studies from the WRA report, which are summarized below, illustrate the benefits of this type of collaboration.

Joint rebates

In 2008, the investor-owned, California-based utility PG&E, along with several water agencies in California, offered a rebate program for high-efficiency clothes washers. The rebate in 2013 ranged from $100 to $125 – this includes a $50 rebate from PG&E and a variable rebate from $50 to $75 from the water utility. PG&E has seen a 63% increase in customer participation since the water utilities joined the program and the water utilities have seen a 30% increase in their customer participation. The program has since expanded to 41 water agencies (municipal, regional and private utilities).

Joint audits

Three Texas utilities – Austin Water Utility, Texas Gas Service and Austin

Energy – collaborated in 2011 to develop a "Multifamily Energy and Water Efficiency Program." The program, funded in part by the US Department of Energy, is designed to conserve water, electricity and gas. The program provides resource efficiency improvements for multifamily residential dwellings and is projected to upgrade approximately 1,900 multifamily units, resulting in approximately 4.7 million kilowatt-hours of energy savings and 10 million gallons of water. One of the key innovations is that the program overcomes the "split incentive" problem where the property owner incurs the cost of the upgrade but the renter earns the resource-efficiency benefits. The holistic approach to resource efficiency created enough benefit to the owners for them to participate in the program and overcome the split incentive problem.

Joint building efficiency upgrades

The Los Angeles Department of Water and Power (LADWP, a municipal utility) and the Southern California Gas (SoCalGas, an investor-owned utility) company launched six residential and commercial energy–water programs in late 2012. This collaboration made available to utility customers several resource-efficiency programs concurrently, including a retro-commissioning program to tune up non-residential building equipment, a program to implement energy- and water-efficiency projects for residential and small business facilities, and energy efficiency upgrades for the Los Angeles Unified School District.

All of these programs demonstrate that there is greater value in combined and/or coordinated resource efficiency programs. Breaking down silos and offering consumers integrated programs to promote resource efficiency benefits the power and water utilities, consumers and the public sector in general.

One of the best locations to examine the connection between water and energy and the need for coordination of conservation and efficiency programs is in the US West and in particular California. In the US West, population centers are often far from water supplies, which require extracting and transporting water long distances. In California, about 19% of electricity use, 32% of all natural gas consumption and 88 million gallons of diesel fuel are related to water.[35] Reducing water withdrawal and consumption has a direct impact on energy use.

Research by the Pacific Institute also provides an overview of the connectivity, barriers to implementing coordinated programs and case studies where these barriers have been overcome. It is worthwhile focusing on the most significant barriers and recommendations for overcoming the barriers to coordinated programs.

The biggest barriers identified by a survey conducted by the Pacific Institute are:

- The water sector has limited or inconsistent funding available to invest in combined programs.

- Limited staff time.

- Insufficient guidance about how to equitably allocate costs and benefits among project partners.

- Water-related pricing policies (e.g., few mechanisms for cost-recovery and concerns about revenue stability).

- Lack of established relationship between potential water and energy partners.

Other notable barriers include fragmentation within and across sectors, a lack of appetite for innovation and risk-taking within the water sector, a lack of directive from regulatory agencies, and a lack of awareness about water–energy connections at the utility level.

The Pacific Institute highlights many of the same utility programs to save water as Western Resource Associates, and notes that successful programs managed to address the following challenges:

- Obtaining funding from multiple sources.

- Offering customers new or expanded services at lower time and money costs than what would have been required to implement such programs individually.

- Utilizing a third party to administer the program.

- Demonstrating the value in connecting efficiency programs – that saving water saves energy.

4.3.2 Rethinking utilities

A 2013 report by the Johnson Foundation titled "Building Resilient Utilities – How Water and Electric Utilities Can Co-Create Their Futures" provides an overview of the issues, challenges and potential solutions to address the nexus of electricity generation and utilities for both water and wastewater. The key point made in the report is that in the United States, water and wastewater utilities are among the largest consumers of electricity. At the same time, US electric utilities are facing business continuity risks from water scarcity. These challenges are driving increased innovation; for the water sectors we are seeing innovation in increased energy efficiency, elimination of net energy use and nutrient recovery and increased water efficiency. Electric utilities are innovating to reduce greenhouse gas emissions, address air quality issues, increased energy efficiency and renewable energy into their portfolios.

However, the report concludes that innovations are "largely occurring independently of one another."

FIGURE 3. Framework for Change from the Johnson Foundation "Building Resilient Utilities" (November 2013).

PHASE I	PHASE II	PHASE III
Optimize Existing Systems	**Transition to More Resilient Systems**	**Implement Transformative Systems**
Phase I includes increasing energy efficiency, reducing physical vulnerability and improving overall management of legacy infrastructure systems where and when possible, within the scope of typical operations. Example actions include: replacing old way system pumps with state-of-the-art, high-efficiency pumps; relocation plant switchboards and generators away from basements and other areas at risk of flooding; installing comprehensive monitoring systems that enable better tracking of system operations and identification of leaks; and coordinating operations and maintenance planning across municipal departments to increase efficiency and reduce costs.	Phase II involves incorporating proven innovations into legacy systems, to move them beyond the scope of typical operations while enhancing adaptive capacity and mitigating climate change where possible. Example actions include: incorporating green infrastructure into municipal stormwater management; linking decentralized water treatment and distribution nodes into the grid; connecting anaerobic treatment processes and/or renewable energy generation to wastewater treatment systems; developing anticipatory post-disaster recovery and rebuilding plans; and leveraging market forces to encourage water use efficiency among utility customers.	Phase III involves seizing opportunities to implement and demonstrate the technology, best management practices and multiple benefits of "new paradigm" water infrastructure systems. Key characteristics of sustainable and resilient new paradigm systems include: right-sized, wisely sited facilities; decentralized, co-located water treatment and energy generation facilities, closed loop systems; resource recovery and water reuse; integrated management of drinking water; wastewater and stormwater; use of triple-bottom-line accounting procedures that consider ecosystem services; full-cost pricing; and robust financing mechanisms.

A "framework for change" was developed as a result of the Johnson Foundation work (Figure 3). This framework recognizes that there is a continuum for change – a need to optimize existing systems, transition to more resilient systems and implement transformative systems.

The Johnson Foundation identified the following areas of opportunity for the energy and water utility sectors to collaborate and lead the way toward the infrastructure of the future together:

- Proactively manage electricity demand.

- Enhance energy efficiency and bolster alternative generation.

- Shift to clean power sources.

- Leverage resource recovery opportunities.

- Shape federal and state policy.

- Engage in community development and resilience planning.

While these programs demonstrate that there is business value in addressing energy and water use cooperatively between electricity utilities and water utilities, the solutions are incomplete. The real question is what does an "integrated utility" of the future look like?

The big ideas and key attributes of an integrated utility of the future from the Johnson Foundation work can be summed up as: "The utility of the future will provide water delivery, wastewater treatment, energy generation and solid waste management services in an integrated way that optimizes the use of all resources and eliminates waste."

Specifically, they envisioned the following attributes of the integrated utility:

- Designed to achieve unprecedented levels of sustainability and resilience.

- Provides water delivery, wastewater treatment, energy generation and solid waste management services.

- Embraces systems thinking and the precautionary principle; exceeding regulatory compliance would be the minimum standard.

- Applies triple-bottom-line analysis in decision-making for all aspects of the facility, with facilities designed to minimize freshwater use, water consumption and GHG emissions to approach or achieve net-zero emissions.

- Sized to meet the needs of customer populations as efficiently as possible.

- Embraces public–private partnerships for infrastructure financing.

- Gives local communities control over some facilities through a co-op business model that supports community values and generates a variety of direct community benefits.

- Serves the "customers of the future" – educated and actively engaged in how the utility is managed.

The National Association of Clean Water Agencies (NACWA), the Water Environment Research Foundation (WERF) and the Water Environment Federation (WEF) the same year crafted their vision for a water resources facility of the future, summarized in the 2013 report, "The Water Resources Utility of the Future. Blueprint for Action, 2013." The work of these organizations echoes the Johnson Foundation conclusions and thinking.

The "Blueprint for Action" members created a vision for the "utility of the future." The utility of the future "transforms itself into a manager of valuable resources, a partner in local economic development, and a member of the watershed community seeking to deliver maximum environmental benefits at the least cost to society."

This transformation involves not only reclaiming and reusing water and finding commercial uses and energy sources from biosolids and liquid streams. The utility of the future is distributed, automated and circular:

- Distributed: facilities will likely be distributed to avoid transporting reused wastewater long distances and thereby saving costs for energy, infrastructure replacement and maintenance.

- Automated: real-time monitoring, Web-enabled mobile devices, and cloud computing will enable unattended operations, adjust processes in real time, communicate with customers, and manage the commercial process.

- Circular: embracing the "circular economy" as resources such as water, nutrients, solids, heat, energy and other constituents will be reused and not discarded.

This vision includes not just greener activities, but also those that are more connected to stakeholders within the watershed. This will mean using natural land-based solutions such as constructed wetlands in place of concrete and steel containment and treatment structures to manage storm water, and collaborating with clean water utilities to implement water quality solutions that save money while preserving natural resources.

Although the utility of the future that meets these standards is still years away, the Blueprint for Action report highlights several utilities that are making significant progress:

- **The East Bay Municipal Utility District (EBMUD)** in Oakland, California, launched a program to blend community food waste – fats, oils, and grease from local restaurants and food waste from wineries and farms – with their own biosolids to generate electricity from methane. This project now creates 55,000 megawatt-hours of energy per year, enough to meet the utility's entire energy demand and send the excess to the grid. It also saves $3 million per year and helped significantly reduce the utility's carbon footprint.

- **Singapore's Public Utility Board (PUC)** has been treating and reusing municipal wastewater to drinking water quality since 2003. Today, 30% of the city-state's water needs are met with reused wastewater produced by its three "NEWater" plants, and by 2060, this will reach 50%.

- **The Australian Government** has invested AU$1.5 billion in the "Water Smart Australia" program to transform how the country's utilities use and manage water resources, in part by encouraging quicker uptake of smart technologies. In one AU$100 million project, two private firms, Veolia Water and AquaNet Sydney, signed a 20-year agreement to supply public utility Sydney Water with about 5 million gallons per day of recycled water. The firms divert wastewater from discharge pipes and use membrane filtering to purify the water before delivering it Sydney Water for use in industrial cooling, irrigation and firefighting.

4.3.3 Decentralized water treatment

One of the more significant trends in water (in some ways mirroring the decentralization of power generation through rooftop solar) is the increased interest in smaller, decentralized wastewater treatment systems. This interest is driven by several factors, including changing regulations, increased urbanization, new technologies, a need for energy-efficient water and wastewater treatment and the need for infrastructure repair coupled with limited funding.[36]

A decentralized system treats and disperses wastewater from individual homes or a cluster of homes at or near the source of the wastewater discharge. In the United States, the National Onsite Wastewater Recycling Association (NOWRA) is the largest organization focused on the decentralized wastewater treatment market.

Decentralized systems can provide services for a cluster of homes, a subdivision or small community as well as commercial and industrial complexes. These systems take advantage of the soil's ability to remove or transform pollutants in effluent as it passes through the soil, thereby avoiding discharges to surface waters and maintaining the quality and quantity of groundwater supplies.

Onsite systems are a promising green technology in part because the treated effluent is used to recharge local aquifers. Decentralized waste-water management systems are now also reusing or recycling treated effluent, which can then be used for irrigation, toilet and urinal flushing or in commercial boilers. All of these uses reduce the demand for scarce potable water and protect and preserve existing water sources.[37]

....

CHAPTER 5

21st-Century Thinking: A New Framework with New Rules

I DON'T BELIEVE IN BUSINESS-AS-USUAL PROJECTIONS. While they are helpful in calling attention to an issue to catalyze change we seldom follow the projections – resource stress and scarcity drives innovation. Innovation kills business-as-usual projections, and innovations can also solve the energy–water–food nexus challenge.

Old ways of innovating are giving way to new thinking and tools to accelerate innovation. One only has to look at crowdsourcing and prize competitions for examples. We now have the tools to mobilize the best minds globally to address a wicked problem such as water scarcity, low carbon energy and access to nutritious foods. The X-Prize's success in tapping into global talent to tackle a big challenge such as accessible space flight is just one example. The X-Prize is now moving into addressing other big challenges, including education and healthcare.

Briefly, in their own words, an X-Prize is "an incentivized prize competition that pushes the limits of what's possible to change the world for the better. It captures the world's imagination and inspires others to reach for similar goals, spurring innovation and accelerating the rate of positive change." The following criteria define an X-Prize challenge:

- Sets a bold and audacious goal.

- Targets market failures.

- Defines the problem vs the solution.

- Is audacious but achievable.

- Is winnable by a small team, in a reasonable time frame.

- Is telegenic and easy to convey.

- Drives investment.

- Provides vision and hope.

If you want an in-depth view of how prize competitions, and in particular the X-Prize, can address wicked problems, read *Abundance: The Future is Better than You Think* by Peter H. Diamandis and Steven Kotler (Free Press, 2012).

Another big idea to tackle wicked problems is the power of innovative collaboration. While prize competitions such as X-Prize drive technology innovation, there is the "soft side" of innovation – catalyzing change through innovative collaboration. In the world of water stewardship this is referred to as collective action and is an essential component in tackling water issues, including access to water, water quality, sanitation and hygiene.

The Solution Revolution by William D. Eggers and Paul Macmillan (Harvard Business Review Press, 2013) lays out one of the big changes in how complex issues are solved: by bringing together a diverse group of stakeholders, including businesses, governments and social enterprises.

No longer is any one group of stakeholders expected to address complex societal and environmental problems. The new model of collaboration is gaining widespread recognition as a smart and effective way forward.

Watch how prize competitions such as the X-Prize and the Solution Revolution can be leveraged to address the energy–water–food nexus challenge.

We are at a time where we can address the projections for water scarcity and provide sustainable energy and food for the current and projected global population. However, it will require accelerating the pace of adoption of innovative technologies and partnerships and public policies. It is within reach – even if it will not be easy. It also means abandoning our business-as-usual mindset and embracing a rethinking of how we have historically managed these resources. Abandoning the old way of thinking and moving to a 21st-century mindset powered by new technologies, collaboration frameworks and public policies is possible.

What we need is a new framework for thinking and new rules – as exemplified by prize competitions and collaboration.

5.1 A new framework

To create this new framework (Figure 4), I grouped initiatives into two major categories – "soft path" and "technology" solutions – similar to the soft path of the Solution Revolution and the hard path of prize competitions to develop technology innovation. We need both a soft path and a technology path, and the two are connected. The framework is meant to focus thinking and illustrate connectivity while not creating new silos of thinking. Soft path solutions consist of collective action and public policy initiatives, while technology solutions consist of connectivity and resource productivity initiatives.

FIGURE 4. A framework for addressing the energy–water–food nexus.

Framework for Addressing the Food Energy Water Nexus

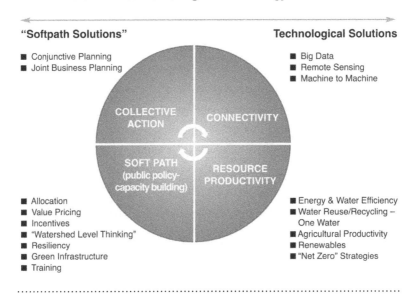

"Softpath Solutions"   Technological Solutions

- Conjunctive Planning
- Joint Business Planning

- Big Data
- Remote Sensing
- Machine to Machine

COLLECTIVE ACTION · CONNECTIVITY · SOFT PATH (public policy-capacity building) · RESOURCE PRODUCTIVITY

- Allocation
- Value Pricing
- Incentives
- "Watershed Level Thinking"
- Resiliency
- Green Infrastructure
- Training

- Energy & Water Efficiency
- Water Reuse/Recycling – One Water
- Agricultural Productivity
- Renewables
- "Net Zero" Strategies

5.2 Soft paths

It is easy to gravitate to technology innovation to address energy, water and food issues. We always look for a technology solution, as they usually don't require behavior changes – the soft path solutions. However, one of the most exciting opportunities resides in collective action or aligned action initiatives. Mobilizing diverse stakeholders to tackle wicked problems is a solution that shows great promise. Technology solutions coupled with public policy changes and collaboration are powerful and together can solve complex problems.

5.2.1 Collective action

Wicked problems such as water, managing food waste or providing energy to emerging economies can only be addressed through collective action frameworks. Collective action as a term is increasingly used in the water sector to address how companies, NGOs and the public sector are rallying to address a range of pressing issues. There is a recognition that no single entity can solve these complex problems – the Solution Revolution described by Eggers and Macmillan.

Collective action programs include industry-specific initiatives, cross-industry initiatives, joint business planning and conjunctive planning. These collective action programs reflect the coming together of stakeholders within an ecosystem or several ecosystems to address challenging issues.

A few examples of collective action initiatives focused on water offer an overview of what is already taking place:

- The UN CEO Water Mandate (**www.unglobalcompact.org/Issues/ Environment/CEO_Water_Mandate/**) and NGOs work with multi-national corporations, and corporations in turn work closely with their supply chain partners. The CEO Water Mandate developed the online Water Action Hub to facilitate partnerships within strategic global watersheds.

- The 2030 Water Resources Group released an online database of case studies to address water scarcity risks. It is designed to facilitate adoption of leading practices to cover a wide range of common scarcity challenges, as well as proven solutions. The group offers a free download of the full catalog of in-depth solutions (**www.waterscarcitysolutions.org**).

- The World Business Council on Sustainable Development (WBCSD; **http://wbcsd.org**), along with member companies, developed WASH at the Workplace (**www.wbcsd.org/washatworkplace.aspx**), a long-term vision and implementation plan to address access to safe water, sanitation and hygiene in the workplace. Among the organizations that have signed the pledge to date are Greif, Nestlé, Borealis AG, Roche Group, Hindustan Construction Company, the Environmental Defense Fund and Unilever.

5.2.2 Public policy

Public policy needs to change – and change fast – to address the "new normal" of water scarcity and impacts to the food and energy sectors. In many instances, public policy changes slowly compared to how multinational companies address business issues.

For example, we are seeing public policy slowly addressing the droughts in the US Southwest. This is resulting in public policy changes in water allocation and conservation, and even a re-examination of agricultural policy – for instance, asking the important question of why subsidize the growing of water-intensive crops in a water-scarce region – as well as a recognition that renewable energy has the added advantage of having a low water footprint.

The changes in public policy for water are taking the form of questioning century-old water rights laws, the relationship between power and water utilities and water use and allocations in the agricultural sector. New funding for water projects, coupled with a renewed call for regulations and policies to address the need for energy and agricultural production, is emerging in such US states as California and Texas.

The water and power utility sectors offer an interesting area for innovation. A move to integrated and smarter operations – which overlaps with the hard-path solutions of technology and resource productivity – is occurring in the United States and elsewhere. In these sectors we are seeing a move toward "conservation synergy" and distributed generation.

5.3 **Technology solutions**

The big opportunities in technology have to do with big data and connectivity (under the maxim "you can't manage what you can't measure"), renewable energy generation, improved water treatment as with low-energy desalination and precision agriculture such as urban agriculture technologies. As some of the most innovative technology solutions are in the agricultural sector, most of the examples provided below are focused on food and water.

5.3.1 Connectivity

Through digital applications the world is increasingly connected and data are available to drive smart decisions regarding the use of resources. We can now collect data remotely (which is being accomplished by the NASA GRACE satellites collecting data on global water resources), in the field through the use of drones[38] and empowering individuals with mobile applications.[39] Machines can now communicate with each other (as in the John Deere example below) and with us (as with mobile phone apps). This connectivity is resulting in the more efficient use of resources and the creation of new business models.

Connectivity includes big data, remote sensing, machine-to-machine communication and digital applications such as social media. The power of connectivity is emerging as a driver for smarter, precision agriculture. Whether it takes the form of traditional agriculture companies buying data and information companies, or agriculture machinery companies embedding smart sensors into their products, a wave of innovation in connectivity is taking place.

Three recent examples show the ways in which connectivity and big data are transforming agriculture:

- In 2013, Monsanto acquired the Climate Corporation in a $930 million deal. Climate Corporation built a business on evaluating rainfall and soil quality data to help farmers predict crop yields.[40] Monsanto also plans to sell Climate Corporation's crop insurance products to farmers internationally. This acquisition comes on the heels of Monsanto's $250 million acquisition of Precision Planting, a company that allows farmers to plant seeds at various depths and spacing to allow for different treatment options. The acquisitions fit into an overall plan the *New York Times* describes as creating "two-way farm machinery systems that took up and receive data, giving farmers better sense of what to plant and how much water and fertilizer to use."

- John Deere is now using sensors in several of its products to increase farm productivity,[41] for instance adding sensors to its latest equipment lines to help farmers manage their fleets and to decrease tractor downtime while also saving on fuel. The information is combined with historical and real-time data on weather prediction, soil conditions, crop features and many other data sets. The company has created a Web-based platform and mobile phone apps to helper farmers figure out which crops to plant where and when, when and where to plough, where the best return will be made with the crops and even which path to follow when plowing. These tools aim to increase the productivity and efficiency of the crops and in the end result in higher production and revenue.

- In December 2013, Trimble acquired the assets of C3 of Madison, Wisconsin to provide farmers with data and decision-based recommendations. C3 combines crop information with detailed soil data to enable a more complete assessment of the site-related factors that impact crop yield, quality and health. C3 products include the Soil Information System, a collection of innovative tools and techniques for digital, 3-D mapping of soil characteristics; PurePixel Vegetation Mapping software, which provides processing of vegetative aerial imaging for crop health analysis; Soil Inventory, which combines multiple precision agriculture technologies to help determine field variability; Agricultural Forensics techniques for processing soil information and vegetative aerial imaging to understand complex issues; and Soil Imaging Penetrometer, a miniature soil-imaging system that views and analyzes on-site soil imagery at high resolution.

5.3.2 Resource productivity

We need to do more with less. The agricultural sector is increasing productivity – less water, energy and nutrients to grow more food – that may result in a second green revolution.

Additional changes are needed in our food, agriculture and trade systems in order to increase diversity on farms, reduce our use of fertilizer and other inputs, support small-scale farmers and create strong local food systems. That is the conclusion of "Trade and Environment Review 2013: Wake Up Before it is Too Late", a new report from the UN Commission on Trade and Development (UNCTAD; **http://unctad.org/en/Pages/Home.aspx**). The report lays out the shift toward more sustainable, resilient agriculture; livestock production and climate change; the importance of research and extension; the role of land use; and the role of reforming global trade rules.

The report links global security and escalating conflicts with the urgent need to transform agriculture toward what UNCTAD calls "ecological intensification." This is a radical change, described in the report as "a rapid and significant shift from conventional, monoculture-based and high-external-input-dependent industrial production toward mosaics of sustainable, regenerative production systems that also considerably improve the productivity of small-scale farmers."

Along with the need for a dramatic transformation in agricultural practices, the UNCTAD report identifies the key areas for stakeholders to focus their efforts:

- Increasing soil carbon content and better integration between crop and livestock production, and increased incorporation of agroforestry and wild vegetation.

- Reduction in greenhouse gas emissions of livestock production.

- Reduction of GHGs through sustainable peatland, forest and grassland management.

- Optimization of organic and inorganic fertilizer use, including through closed nutrient cycles in agriculture.

- Reduction of waste throughout the food chains.

- Changing dietary patterns toward climate-friendly food consumption.

- Reform of the international trade regime for food and agriculture.

On that final point, the report authors highlight the impacts trade liberalization has had on agriculture systems and push for agreements that promote the development of locally based, agroecological systems that better support farmers.

CHAPTER 6

Closing Thoughts: Abandon Business as Usual

THE PROJECTIONS OF WATER, FOOD AND ENERGY needs for a global population of 9 billion are not encouraging. However, these are projections of business as usual. While business-as-usual scenarios are helpful in mobilizing action to address complex environmental and social problems, the scenarios are not a forgone conclusion.

Innovations in technology, business models and even public policy have the potential to chart a better path forward. The potential is evident – what is required is the will to address these challenges together by recognizing that 9 billion people deserve access to energy, food, safe water, sanitation and hygiene to ensure a reasonable quality of life.

❝The best way to predict the future is to create it.❞ PETER DRUCKER

....

References

1. http://www.apsc.gov.au/publications-and-media/archive/publications-archive/tackling-wicked-problems

2. http://www.wbcsd.org/Pages/EDocument/EDocumentDetails.aspx?ID=15994

3. http://water.usgs.gov/GIS/huc.html

4. http://www.arb.ca.gov/fuels/lcfs/workgroups/lcfssustain/hanson.pdf

5. http://pubs.usgs.gov/fs/2009/3098/pdf/2009-3098.pdf

6. http://pubs.usgs.gov/circ/1405/

7. http://www.prb.org/pdf09/09wpds_eng.pdf

8. http://www.bp.com/content/dam/bp/pdf/statistical-review/BP_World_Energy_Outlook_booklet_2013.pdf

9. http://www.fao.org/docrep/016/ap106e/ap106e.pdf

10. http://iiesi.org/assets/pdfs/copenhagen_bellet.pdf

11. Zilberman, David. Department of Agricultural and Resource Economics, University of California at Berkeley (Spring 1999).

12. http://water.columbia.edu/files/2012/08/PunjabWhitepaper-Final-Version-2-July-20121.pdf

13. https://www.gov.uk/government/uploads/system/uploads/attachment_data/file/69565/pb13786-biofuels-food-security-120622.pdf

14. http://www.nrel.gov/docs/fy11osti/50900.pdf

15. http://www.gracelinks.org/1828/food-water-and-energy-know-the-nexus

REFERENCES

16. http://www.tva.com/environment/reports/bfn_towers/BFN_CoolingTowers_
FEA_2010-53_Final2.pdf

17. http://dx.doi.org/10.1175/2010JCLI3349.1

18. http://www.theguardian.com/global-development-professionals-network/
2014/jan/30/ethiopia-renewable-energy-project

19. https://www.youtube.com/watch?v=ctjz8siAOGU

20. http://www.water-energy-food.org/documents/understanding_the_nexus.pdf

21. http://www.odi.org/publications/7441-biofuels-land-agriculture-indonesia-
ethiopia-zambia-mozambique-tanzania#downloads

22. UN World Food Programme: http://v.gd/kUPJe8

23. http://www.nytimes.com/2014/10/12/world/dam-rising-in-ethiopia-stirs-
hope-and-tension.html

24. http://www.bloombergview.com/articles/2011-07-06/why-we-care-about-
the-price-of-water-in-china-peter-orszag

25. http://www.waterfootprint.org/Reports/Liu-Savenije-2008-China.pdf

26. http://www.water-technology.net/projects/south_north/

27. http://cubeimg.zhongsou.com/1/1218043628648503218178755.pdf

28. http://www.wsp.org/sites/wsp.org/files/publications/Water-and-Energy-
World-Bank-SIWI2013.pdf

29. http://www.npconline.co.za/MediaLib/Downloads/Home/Tabs/Diagnostic/
MaterialConditions2/Water%20security%20in%20South%20Africa.pdf

30. http://www.plosone.org/article/info%3Adoi%2F10.1371%2Fjournal.pone.
0007940

31. http://www.sfenvironment.org/article/recycling-and-composting/mandatory-
recycling-and-composting-ordinance

32. http://www.scientificamerican.com/article/more-food-less-energy/

33. http://www.forbes.com/sites/bethhoffman/2012/04/06/food-waste-visualized-infographic-1-of-2/

34. http://www.fao.org/ag/ca/CA-Publications/17_ZIMBABWE_LR.pdf

35. http://www.pacinst.org/wp-content/uploads/2013/09/pacinst-water-energy-synergies-full-report.pdf

36. http://www.wateronline.com/doc/economy-drive-trend-toward-decentralized-water-treatment-0001

37. http://www.nowra.org/content.asp?contentid=35

38. inniribo2015, "The Robot Report, Are Ag Robots Ready?", 18 November 2014.

39. http://www.theguardian.com/sustainable-business/2014/aug/27/mobile-apps-california-drought-uk-floods-india-drinking-water

40. "Why Big Ag Likes Big Data," New York Times, 2 October 2013, http://bits.blogs.nytimes.com/2013/10/02/why-big-ag-likes-big-data

41. http://www.bigdata-startups.com/BigData-startup/john-deere-revolutionizing-farming-big-data/

....

For Product Safety Concerns and Information please contact our EU
representative GPSR@taylorandfrancis.com
Taylor & Francis Verlag GmbH, Kaufingerstraße 24, 80331 München, Germany

www.ingramcontent.com/pod-product-compliance
Ingram Content Group UK Ltd.
Pitfield, Milton Keynes, MK11 3LW, UK
UKHW040928180425
457613UK00011B/289